"It is one of the greatest paradoxes of recorded history that one of the very same public health measures primarily responsible for the great increase in statistical life expectancy in the Western world should also unsuspectedly be responsible for many of the chronic disorders of later life."

Dr. Joseph M. Price has long held that the chlorine in our drinking water—not the cholesterol in our food—is the major cause of coronary disease.

In this courageous book he challenges the medical establishment with his own bold theory, a thorough and thoroughly frightening examination of the water we drink —that is taking our lives!

CORONARIES/
CHOLESTEROL/
CHLORINE

JOSEPH M. PRICE, M.D.

JOVE BOOKS, NEW YORK

This Jove book contains the complete
text of the original hardcover edition.
It has been completely reset in a typeface
designed for easy reading and was printed
from new film.

CORONARIES/CHOLESTEROL/CHLORINE

A Jove Book / published by arrangement with
Alta Enterprises, Inc.

PRINTING HISTORY
Alta Enterprises, Inc. edition published 1969
Five previous paperback printings
Jove edition / March 1981

ISBN: 0-515-09461-7

Jove Books are published by The Berkley Publishing Group,
200 Madison Avenue, New York, New York 10016.
The name "JOVE" and the "J" logo
are trademarks belonging to Jove Publications, Inc.

PRINTED IN THE UNITED STATES OF AMERICA

20   19   18   17   16   15

CONTENTS

5

PREFACE

"Give not that which is holy unto the dogs, neither cast ye your pearls before swine, lest they trample them under their feet, and turn again and rend you."

- MATT. 7:6 -

The original manuscript of this book was completed in May, 1967. Shortly thereafter, and before arrangements for publication could be made, the author was drafted into the Medical Corps, U.S. Army with orders for shipment to the combat zone in Vietnam.

While the author was serving in an infantry battalion in Vietnam, the basic concepts revealed in this book were released to the nationwide news services in hope that progress in implementation of the proposed practical suggestions could be forthcoming without delay.

Regrettably, the establishment has seen fit to seemingly ignore this amazing health breakthrough. It would be reasonable to think that the immediate result of the release of this chlorine-

heart attack concept would be to evoke condem-
nation, or at least statements of doubt. But one
should not underestimate the cleverness of "the
powers that be". To argue against a new concept
results in publicity which in turn evokes demands
for clarification—and then only the truth can
win. Much more effective is the conspiracy of
silence.

An agent who contracted to promote dissem-
ination of the theory suddenly dropped out, al-
luding vaguely to resistance by chemical firms. A
lawyer reported that the Commonwealth of
Pennsylvania seems ready to allow ozonation of
water supplies, presumably rather than permit
the author's appearance in court to testify in
respect to his chlorine-coronary heart attack
theory. An interested Congressman relayed a
fairly long synopsis with answers to questions
posed by researchers of the U.S. Public Health
Service, upon request, to the Surgeon General—
with no further developments. And so on.

A conspiracy of silence . . . ?

Two years have passed—and one *million* more
Americans have died of heart attacks alone. In
view of this it was decided to publish the complete book for general consumption. Finally you,
the reader, can have all of the facts and make
your own educated decisions on the validity of
my concepts and conclusions especially as they
apply directly to you.

NOTE

On perusal of this book, the reader will soon note a somewhat unusual style of writing. The author was primarily concerned with getting fairly complex medical concepts across to the non-medically oriented reader. Hence the catechismal questions and answers approach and repeated review of important ideas and relationships. Getting across the message was considered vastly more important than literary "style".

It is interesting to note that although two years have passed since the basic manuscript of this book was completed only very minor changes were necessary in preparing it for publication. All new studies and observations made in this intervening period have only substantiated my findings and conclusions.

In addition, as of the present no physician or medical researcher has presented any reasonable arguments against my work. This should speak for itself . . .

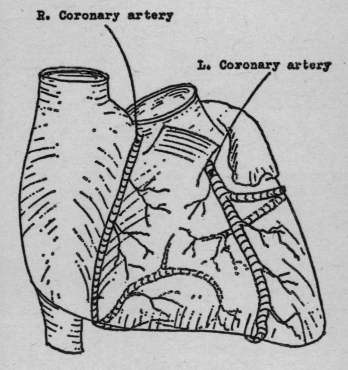

THE CORONARY ARTERIES

CHAPTER 1

A FULL SCALE EPIDEMIC

The United States is in the midst of an unprecedented disease epidemic. The word "epidemic" brings to mind horrifying images of infectious diseases sweeping a nation. This situation obviously does not exist in the U.S. today. Nevertheless, each year heart attacks and strokes kill and maim many times the number of persons afflicted by all the serious infectious diseases combined.

Just about everyone knows somebody, relative or friend, who has had or died from a heart attack or stroke. These two disease entities, along with cancer, are so widespread and of such importance in America today that the Congress of the United States has appropriated many millions of dollars to set up nation-wide regional health centers devoted exclusively to medical research on and treatment of these afflictions.

This constitutes the largest and most ambitious medical project, in terms of manpower, facilities, and expense, ever attempted in recorded history.

Cardiovascular and cerebrovascular diseases ("coronary heart disease and stroke") together constitute the main as yet uncontrolled disease processes afflicting mankind today.

The coronary heart attack (the same disease process is known under several other names, such as "coronary thrombosis" or "myocardial infarction", but in this presentation we will generally use the term "heart attack") is by a substantial margin the number-one killer in the United States today. More than one-half million Americans die each year of heart attacks. But the true size of the problem is not properly conveyed with such black-and-white statistics. In addition to deaths, one must not forget the vast toll of non-fatal heart attacks—the costs of medical care, loss of wages, the incalculable physical and mental anguish of the patient and his family, etc. Even if the victim does not die the effects on himself and his loved ones may be absolutely devastating nonetheless. From a broader point of view the loss of productivity for survivors is tremendous—a loss for our nation as a whole. Most frighteningly, for certain groups the death rate from this cause is still rising despite massive medical research programs! It is in all truthfulness a full scale epidemic.

CHAPTER 2

THE HEART ATTACK

Especially if you are a middle-aged man, you are probably a bit more than a little concerned about this problem. What really *is* a heart attack? You may be wondering in the back of your mind what it's like to have a heart attack. More to the point, what are the chances that *I* am going to have a heart attack; and hopefully, is there anything I can do to help protect myself against suffering one?

Let us consider these questions one at a time.

Everyone knows that a person who has had a heart attack has had something bad suddenly happen to his heart and as a result he may die. But for the reader to really understand what is going to be explained later regarding the final solution to this killer, a little preliminary explanation is in order regarding what really happens *in the heart.*

First of all, it is necessary to become acquainted at least superficially with a few facts about the anatomy of the heart. The heart is an amazing hollow muscle about the size of your clenched fist which serves the life-sustaining function of pumping blood to all parts of the body. It starts beating a few weeks after conception while the child is still in its mother's womb and continues unabated until death. The heart, like any other muscle in the body, must constantly receive a flow of blood to it carrying food and oxygen in order to live. What is not understood by the vast majority of non-medical persons is that the heart muscle itself cannot use any of the blood contained within its pumping chambers. Instead the supply of blood to the heart muscle comes entirely from two little vessels, called coronary (heart) arteries, which arise from the main artery of the body, the aorta, run along the outside surface of the heart and then enter into the heart muscle to feed it. (See diagram above.)

Since these two little arteries are the *only* source of blood to the heart muscle, when one of them (or one of their branches) suddenly becomes blocked off the portion of the heart muscle supplied by the involved artery dies. This death of a portion of the heart muscle is called in medical terminology a "myocardial infarction" or in common language a "heart attack". If the part of the heart muscle which dies is

16

small the dead portion will be gradually replaced by scar tissue and the patient is likely to recover. If a large part of the heart muscle is affected death may come within minutes or hours; or be the result of complications manifest most commonly during the first two weeks of convalescence.

Now that we understand, at least in general terms, what happens in the heart itself during a heart attack let us consider what happens to the victim—what does he feel and what is it like to have a heart attack? The person who is having a heart attack usually experiences the sudden onset of a severe crushing, compressing type of chest pain which may shoot down the arm and sometimes up the neck, especially on the left side. This pain is often accompanied by a feeling of imminent disaster with associated cold sweat, nausea, weakness and sometimes shortness of breath. Many victims of heart attacks give a history of intermittent chest pains beginning a few days or more before the acute infarct (premonitory pain). A physical examination and electrocardiogram done by a physician at this time usually will *not* reveal significant abnormalities. After the actual heart attack has taken place (chest pain, etc. as above) electrocardiograms and certain blood tests (enzyme studies) ordered by the physician can be used to confirm the diagnosis and in some instances will give an indication of the probable prognosis, i.e. chance of survival.

Now that we have an idea about what it is like to have a heart attack and understand that it is due to sudden blocking off of one of the little arteries carrying blood to the heart muscle, the next question that arises is what causes the sudden occlusion ("blocking off")? While the actual total occlusion of a coronary artery which results in an acute heart attack takes place suddenly, it only occurs in previously diseased arteries affected by a pathological process known as *atherosclerosis*. Atherosclerosis is a particular kind of "hardening of the arteries" characterized by the gradual accumulation of certain fatty substances embedded in the inner wall of the arteries, slowly making the lumen ("opening") smaller and smaller. This process can be compared for the sake of clarity to the gradual deposition of lime scales on the inner surface of hot water pipes, except that in the arteries the process is further complicated by the body's reaction to the deposit of foreign material. The deposits of fatty material which result from this process are called *atheromas* and appear grossly as raised yellow-colored plaques. A heart attack occurs when spontaneous bleeding in or around an atheromatous plaque causes a clot to form which then blocks the blood flow through a coronary artery. (Hence the term "coronary thrombosis".) There could not be a heart attack if the coronary arteries were not partially closed by atheromatous plaques.

You are probably wondering if there is any way to tell if your own coronary arteries are in good condition; or are they partially occluded with atherosclerotic plaques and liable to sudden occlusion with resulting heart attack? The answer is that there is no way to prove that your arteries are definitely in good condition. A person may look and feel fine, his physician may pronounce him physically sound after performing a thorough examination including an electrocardiogram, only to suddenly collapse with an acute heart attack. On the other hand, angina pectoris, a syndrome of intermittent chest pains usually precipitated by exercise, emotional stress, or other factors such as exposure to cold, is *prima-facie* evidence of coronary insufficiency, i.e. the coronary arteries are so narrowed by atherosclerosis that the heart does not receive enough blood to support its functions during certain periods of greater need. So while a man may have advanced coronary atherosclerosis and be "ripe" for a heart attack without any symptoms whatsoever, the presence of angina pectoris usually means that there are quite advanced changes present in the arteries and a heart attack is a definite, everpresent threat.

With reference to the asymptomatic group, there are a series of factors which are correlated with a predisposition to heart attacks. These include:

1. Male sex
2. Cigarette smoking
3. Elevated blood pressure
4. Middle age or older
5. High level of blood lipids (fats)
6. Diabetes mellitus
7. Obesity
8. Low levels of physical activity
9. Family history of heart disease
10. Shortness of stature
11. "Soft" drinking water
12. Others

If you live in America or Europe, the more items on this list which apply to you the more likely you are to be a prime candidate for a heart attack.

You will notice that number one on the list is "male sex". The typical heart attack victim is an overweight, middle-aged man who is physically inactive and smokes cigarettes. But this stereotype leaves us with several important questions unanswered. For example, do women get heart attacks? Yes, but only after the menopause ("change of life"). It is an as yet unexplained fact that premenopausal women are almost totally immune to heart attacks unless they already have certain serious diseases, usually high blood pressure or diabetes. This "immunity" is thought to be related in some way to secretion of estrogens (female sex hormones) and ceases to exist when the ovaries (female sex glands) are removed or stop active functioning. In fact, administration of synthetic estrogens to men with strong predispositions to heart attacks has been suggested, but has failed to gain wide acceptance for obvious reasons.

Do young men or even children ever get heart attacks? Surprisingly enough it is a misconception to think of the heart attack as a disease only of middle-aged and old men. While it is true that the disease is seen most often in these groups, especially in recent years heart attacks have become more common in the younger age groups. We are now seeing not uncommonly otherwise normal men in their early 30's dying of heart attacks, and under special circumstances even fatal heart attacks in children. As will be described in some detail later the underlying disease process of atherosclerosis has been found to oftentimes start in early childhood.

By now it should be clear that a heart attack is but the end result of a more general disease process in which deposits are built up in the insides of arteries. It only seems logical that this buildup does not occur suddenly, e.g. overnight. Does it take months? Years? The facts are that while an actual heart attack is one of the most dramatic events in medicine, the underlying disease process of atherosclerosis is quite the opposite. The infiltration of the artery walls begins in most cases at least 10-20 years before any overt symptoms are evident. But this should not be taken to mean that everyone of a particular age group has arteries which are in the same condition. For instance, some men die at age 45 of heart attacks while other men of identical age have youthful, clean arteries good for another 45 years. Some very old people die with but absolutely minimal evidence of arterial disease.

CHAPTER 3

STROKES

Thus far we have centered our attention on the coronary heart attack. As I hope is quite clear by now, when atherosclerosis involves the coronary (heart) arteries it gives rise to heart attacks. But atherosclerosis, that form of "hardening of the arteries" which results from the building-up of fatty deposits on the inside of arteries, may occur almost anywhere in the body. As a general rule it may be said that the signs and symptoms resulting from an advanced atherosclerotic process involving an artery are the signs and symptoms of lack of blood supply to the particular organ(s) supplied.

When the arteries feeding the brain are affected by atherosclerosis the stage is set for the occurrence of that other great killer and maimer of modern civilization, the stroke. A stroke occurs when the blood supply to a part of the brain is compromised. With loss of a source of free-flowing blood to nourish it, the affected part of the brain may die. Unfortunately, in contrast to the heart which is fairly unspecialized and can "make up" for heart tissue destroyed by a heart attack with surviving tissue, brain tissue tends to be highly specialized and therefore when certain parts of the brain die functions related to such areas are oftentimes lost forever—hence the paralyses, loss of ability to talk, read, or to understand speech or the written word, etc. following strokes. The two great subdivisions of stroke are the cerebral hemorrhage (bleeding from a vessel into the brain substance) and the cerebral thrombosis (a blood clot blocking a cerebral-brain vessel which has been narrowed by plaque). With a few exceptions (e.g. congenital berry aneurysm) the great majority of strokes have the process of atherosclerosis as the underlying cause—either by weakening artery walls with consequent rupture and bleeding into the brain or by narrowing the lumen of arteries in the brain predisposing to sudden occlusion by a blood clot. The so-called "small stroke" most commonly results from the breaking off of small portions of atheromas and their blocking off of small cerebral vessels.

CHAPTER 4

OTHER ATHEROSCLEROTIC DISEASE ENTITIES

Although heart attack and stroke are the two immensely important end results of atherosclerosis, there are several other clinical entities secondary to the same process. For instance if the arteries supplying the brain are progressively narrowed due to growing atheromas, but without an acute episode of occlusion or bleeding which would cause a stroke, there is evidenced signs of chronic cerebrovascular insufficiency—spells of dizziness, transient paralyses, forgetfulness and loss of mental acuity. The commonly seen syndrome of senility of the aged is basically a result of chronic vascular insufficiency secondary to atherosclerosis. If instead the atherosclerosis affects mainly the arteries to the legs we see evident the syndromes related to peripheral artery disease—intermittent claudication (pain in the leg muscles on walking but relieved by rest, once again due to vascular insufficiency) or gangrene of the toes, feet or even legs. Aneurysms of the abdominal aorta (great dilatations of part of the body's main artery which may burst resulting in sudden death from hemorrhage) are almost all secondary to weakening of the vessel walls from atheroma formation. Finally we should not forget that the vast majority of cases of physiologic sexual impotency in the male is due to atherosclerosis blocking off the normal blood supply to the sexual organs. A psychologically normal man who is free from atherosclerosis and serious neurological disorders should be sexually potent well into his 70's or 80's! There are a few other less common disorders and manifestations related to atherosclerosis which are not of sufficient prevalence or importance to discuss at this time.

From these brief descriptions it can be realized that atherosclerosis is the basic pathological process behind many of the most devastating diseases of today. In particular it is the underlying cause of the majority of the chronic diseases of modern civilization which are ravaging the land and against which present-day medical science can do but little.

To summarize, atherosclerosis is the basis for:

1. *Heart attacks* and angina pectoris
2. *Strokes*
3. Peripheral arterial disease, including in-

termittent claudication and gangrene of the extremities

4. Aneurysms of the abdominal aorta and other arteries
5. Senility of old age
6. Male sexual impotency due to vascular insufficiency
7. Others

CHAPTER 5

THE CHOLESTEROL THEORY

The next consideration which quite naturally arises is what exactly *is* this special form of "hardening of the arteries" called atherosclerosis? Is it known why the arteries get clogged up with fatty deposits? Let me here and now make the flat statement that modern medical science does *not* acknowledge a definite known causal factor for atherosclerosis. In fact there are strong disagreements among medical scientists as to such basic knowledge as the exact mechanism of accumulation of lipids in the inner linings of arteries, much less the underlying cause of the same. The only things that are known as to the cause of atherosclerosis are certain *correlations*, and a correlation does *not* imply causation. (A correlation, in simple terms, means that two different things tend to exist together but does not necessarily mean that one causes the other. As an absurd example, I am told that there can be demonstrated a close correlation between the amount of bananas imported into England and the birth rate there. Less facetiously, and of some importance because it demonstrates a significant point regarding correlations between atherosclerosis and various factors—as will be explained in more detail later—is the fantastic correlation between number of telephones per unit of population and the cardiac (heart) death rate. This correlation is more striking than the correlation between atherosclerosis heart attacks and *any* of the dietary factors!)

While it has been acknowledged that a correlation exists between the development of atherosclerosis and such varied factors as age, sex, genetic constitution, endocrine balance, psychic state, drugs, exercise, occupation and climate, most interest and research in the field of the study of atherogenesis (the cause of atherosclerosis) has centered around a fatty chemical substance called *cholesterol*. Cholesterol is a fatty ("lipid") substance which occurs in all animal cells and is essential to life. It is needed by the body as the chemical starting point for many essential compounds. For example, the vital steroidal hormones, which include the basic male and female sex hormones, can be made only from cholesterol. The human brain itself contains a very high percentage of cholesterol.

The body's cholesterol comes from two

sources. First of all the body makes a considerable amount itself from simpler compounds. Secondly, a certain amount of cholesterol is taken into the body whenever certain foods of animal origin (e.g. fatty meats, milk, eggs, butter, etc.) are eaten.

There are several reasons why modern medical researchers have centered most of their heart research investigations around the substance cholesterol, and have assigned it such importance—many medical scientists point an accusing finger at cholesterol as *the* primary causative agent in atherogenesis. First, it has been known for many decades that atheromatous plaques wherever they form are composed mainly of cholesterol. In fact, some claim that an atheroma is essentially an inflammatory response of the body to a cholesterol deposit. Secondly, lesions very similar to human atheromas may be produced at will in certain species of animals by feeding them diets high in cholesterol. Finally, it is thought that there is a correlation between blood cholesterol levels in men and their likelihood of developing one of the atheromatous diseases, e.g. heart attack or stroke. In certain families there is a rare genetic (inherited) defect which results in affected members having fantastically high blood cholesterol levels. Children in these families have been known to die from heart attacks before they reach eight years of age.

It would seem that if cholesterol is the primary cause of atherosclerosis one could avoid heart attacks and strokes by avoiding foods high in cholesterol. Is this a valid statement? It is true that this point of view has many influential proponents. The most widely accepted theory holds that it is the excessive consumption of foods containing cholesterol that predisposes to atherosclerosis and its sequelae. As a result of the widespread acceptance of this cholesterol theory there has been an informal nation-wide campaign on for the last few years to discourage the "excessive" consumption of foods high in fats and cholesterol. As a corollary of this, researchers have discovered that certain types of fat, called "polyunsaturated" and found primarily in vegetable oils, have a vague antagonistic effect in the body on the "saturated" animal fats which are high in cholesterol. Witness the commercialization of this finding in ads and on TV—"So-and-so margarine, highest in polyunsaturated fats!"

Many experts in the field believe that if by diet, drugs or by other means one can lower his blood cholesterol levels he has significantly reduced his chances of developing a heart attack or stroke. Recently medical researchers have been experimenting with certain new drugs which lower the level of blood lipids (fatty substances) including cholesterol in the hope of delaying or eliminating heart attacks and strokes.

CHAPTER 6

IS THE CHOLESTEROL THEORY VALID?

"Well-organized ignorance often passes, unfortunately, for wisdom."

-ANON.-

In spite of the remarks made in the previous chapter, this belief in the possible avoidance of atherosclerosis by dietary manipulation is by no means universal. It has been emphasized by a minority of researchers that *no* direct evidence exists to prove that lowering of blood cholesterol by any method possible will decrease the risk of coronary heart disease or stroke, or even affect the underlying atherosclerosis. In fact, the U.S. Food and Drug Administration has taken this stand in its regulation of advertising to physicians of the new serum lipid (cholesterol and triglyceride) lowering drugs.

Although a causal relationship between cholesterol and heart attacks and strokes (the cholesterol theory of atherosclerosis) has not been proven, the circumstantial evidence may seem pretty convincing. Is there any evidence for the other view, i.e. the cholesterol theory may not be the complete story? Indeed there is. In discussing this other side of the picture it should be remembered that modern medical science hates to admit that it does not know the answer to a problem. So it proposes a hypothetical solution —really an educated guess. Fair enough—one needs a starting point from which to proceed. Unfortunately, for lack of anything better the cholesterol hypothesis has attained almost the status of an established and definitely proven theory—not because anything has been proven but rather because it fills what would otherwise be a scientific vacuum.

The cholesterol theory has become so thoroughly accepted as God-given truth that many otherwise intelligent, scientifically-oriented, insightful medical authorities explain away in a most illogical manner any evidence which would cast a shadow of a doubt on its "unquestionable" validity. The reason for this appallingly unscientific attitude is that rejection of the cholesterol theory would require the substitution of a presumably more acceptable hypothesis. Needless to say, the cholesterol theory, imperfect as it may be, is acknowledged to be the only possibly plausible one presented so far. If it is discarded and not replaced by something more reasonable, medicine would be admitting total impotency against modern mankind's major killers. This, as implied before, is completely unacceptable.

The considerable evidence against the acceptance of the cholesterol theory as the final word on the subject will now be discussed in some detail. The following considerations, I'm sure you'll agree, are quite convincing in themselves.

If all human beings in the world were, and always had been, uniformly subject to heart attacks and strokes with the same prevalence as found in the United States today the hope of finding a definitive solution to the problem of atherosclerosis would be dim indeed. For if the above were true we would probably have to accept atherosclerosis as an inevitable part of the aging process. Fortunately this is not the case at all. Even staunch proponents of the classical cholesterol theory realize that atherosclerosis is a *disease* process which is by no means universal and at least theoretically capable of being retarded and possibly even reversed.

Today in the United States men outnumber women in total heart disease rates by more than one-third; until approximately 1930 the rates were about the same. The only logical explanation is that since about 1930 coronary heart disease has been affecting men, who before then had been almost as immune as premenopausal women are. That is to say that the heart attack, while occurring before 1930, only reached sufficient proportions to affect statistical tables about then. This would tend to imply that some environmental consideration of supreme importance in the etiology (causation) of atherosclerosis began to affect men in the early part of this century. (Remember that 10-20 years of increasing atherosclerosis precedes clinical manifestations.) This could not be cholesterol because cholesterol has been with man as long as there has been man.

I have just introduced the critically important concept that coronary heart disease and other manifestations of atherosclerosis were essentially unknown before the present century, and therefore these disease processes must have an underlying causal agent of rather modern origin. Are there any other historical facts which support this contention? Angina pectoris, the intermittent chest pains which usually imply narrowing of the coronary (heart) arteries by atherosclerosis and often-times precede full-blown heart attacks, was first described only as recently as 1768. And not only was it an exceedingly rare disease for the century and a half following this description but the cases described during this period were not necessarily due to an underlying atherosclerotic process—even today it is an indisputable fact that angina pectoris may rarely be the result of some other disease process such as syphilitic involvement of the base of the aorta or anemia of any origin. These non-atheromatous instances of angina pectoris are exceedingly rare, but then so was the syndrome of angina pectoris itself until some years into this century.

Now let us consider the heart attack itself. The coronary heart attack, which is almost exclusively the end result of atherosclerosis of the coronary (heart) arteries, was completely unknown until early in this century. Hard to believe? Yes, it sure is. I'll be the first to admit that most present-day physicians will immediately challenge this statement. But the pure and simple facts are that the first clinical description of coronary thrombosis (another term essentially synonymous with "heart attack") was made as recently as 1912; the great Canadian-American physician Sir William Osler did not mention the existence of the entity in lectures on heart diseases in 1910; and most amazingly the world famous heart specialist Dr. Paul Dudley White who treated President Eisenhower for his heart attack in the early 1950's did not see his *first* case of myocardial infarction (once again another term essentially synonymous with "heart attack") until after 1920!

Now that I have presented you with historical facts (which may be verified in any good medical library) debunking the concept that atherosclerosis and its clinical manifestations such as the heart attack and stroke are as old as mankind, what is there to say about the role of cholesterol?

Let me make the statement that while cholesterol may be one of many contributing factors to the development of atherosclerosis and its complications, it certainly is not sufficient alone—it is not *the* factor. The evidence for this point of view is substantial. We should first consider a finding closely related to the facts just described: through careful investigation of the literature it has been shown that in England at the end of the 19th century almost one-third of the population consumed dietary fats in amounts which must be considered excessive by present-day standards, and yet heart attacks and other evidences of atherosclerosis were non-existent! Another most impressive finding is that atherosclerotic heart disease has always been unknown in China—700,000,000 people and no heart attacks! And don't let anyone explain this away by saying that it proves that cholesterol *is* all-important because the Chinese consume very little fat. While the peasants have always lived on basically vegetarian diets, many of the traditional dishes of the higher social classes who have been able to afford them are nauseatingly fatty. Yet atherosclerosis has been as non-existent among the well-to-do classes as with the Chinese peasants.

While atherosclerosis and its consequences are practically non-existent among most primitive peoples of the world, there is no better example of high dietary fat intake coupled with the absence of atherosclerosis than seen among the Eskimos. The dietary fat intake of the Eskimos is simply hard to believe—a single adult may eat several pounds of blubber (about as saturated a fat as exists) at a sitting. This fantastic dietary pattern is followed for a lifetime—and yet no heart attacks or strokes from atherosclerosis. If there were no other evidence than this, any thinking man would still question the cholesterol (dietary fat) theory. Nevertheless, there is still considerably more evidence to raise doubts in our minds.

For instance, if you want something a little nearer to home we have that too. A few years ago there was an article in a popular magazine about a small town called Roseto in the hills of Pennsylvania. The people in this town, of Italian descent, tended to be obese and ate a diet abnormally high in animal fats and yet seemed to be immune to heart attacks as long as they did not move out of the community. A little more food for thought . . .

For those with a more classical medical research bent, one should mention the existence of the gerbil, a tiny Mongolian rodent who despite a fat-rich diet and high levels of blood cholesterol shows no tendency toward atherosclerosis.

A number of other similar findings exist; and several more of these will be mentioned when the explanation for all this is elucidated.

CHAPTER 7

THE ANSWER—CHLORINE!

I have so far explained how heart attacks and strokes as well as certain other disorders are due to an underlying disease process called atherosclerosis wherein fatty materials are deposited in the inner lining of arteries, plugging them up. I presented the cholesterol theory of atherogenesis (atherosclerosis being related to high blood levels of cholesterol and possibly other fats which in turn are contributed to by high dietary intake of cholesterol and "saturated" fats) only to reveal convincing evidence that this is not the whole answer. If the cholesterol theory is not the full answer, then one can*not* protect oneself against heart attacks and strokes by avoiding dietary intake of cholesterol and other fats?

This is correct. I am *not* saying that cholesterol has no role whatsoever in atherogenesis and that therefore dietary habits make no difference at all. Cholesterol is one of *many* contributing factors influencing atherogenesis and therefore dietary change *may* affect the disease process to at least a small degree, just as other factors like cigarette smoking, physical exercise, etc. may also have some influence on the matter. I *am* implying that even if a person changes his diet to one low in cholesterol and "saturated" fats he is not guaranteeing himself immunity against heart attack and stroke. Persons who do not smoke get heart attacks and strokes (although admittedly less frequently than those who do). The same statement is true for those who exercise regularly. And I am saying that the same is true for persons who would change their dietary habits.

By now I should imagine that you, the reader, have become a little uneasy. I have told you how heart attacks and strokes are the end result of a disease process called atherosclerosis. I went on to present the classical cholesterol theory of atherogenesis including the possibility of retarding the disease process by changing one's diet. But then I proceeded to produce some rather convincing evidence which would tend not only to discredit the theory but also destroy the one ray of hope with regards to prevention of the disease process by dietary manipulation. You are probably wondering where do we go from here.

Fortunately, this book was not written merely

to give the *coup de grace* to the untenable cholesterol theory. Have I come up with some original theory that all the best medical minds in the entire world over a period of many decades have overlooked?

Indeed I have.

And there are irrefutable scientific *proofs*, reproducible by anyone of average intelligence or more, to back me up.

In retrospect it is amazing how many separate observations point to the definitive solution of the problem and yet have been ignored. It is truly a tragedy that modern medical science in desperation has grasped onto the cholesterol theory (even to the point of ignoring and denying proven facts which would tend to discredit it) to the exclusion of all else. It is just inconceivable to most medical men that such incredibly important, widespread and basically untreatable diseases as heart attack and common stroke could have been essentially unknown less than 75 years ago. If this is true (as it is) it would mean that something has changed in the last 6-7 decades of human history but medical science has failed to see the light. In all fairness let me say that the reason for this shortsightedness will seem much more justified when it is realized exactly what the culprit is.

As implied in the above paragraph, the great stumbling block which has inhibited rational, productive thinking about atherosclerosis was the unreasoning acceptance of the atheromatous diseases as being as old as mankind itself. Instead, as pointed out in the previous chapters, the atheromatous diseases (primarily heart attacks and strokes) are not only less than 75 years old but even today are almost entirely confined to peoples under the influence of modern Western civilization. So therefore we must look for some product of modern Western civilization as *the* cause.

We will now consider a few general causes of disease, with special reference to the atheromatous diseases.

The first thing that always seems to be brought up today is an ill-defined entity called "stress". In fact entire books have been written on this subject alone. Although I know it will bring great cries of distress from certain quarters let me dismiss this concept summarily by saying that if you think that you are living under stressful conditions today, what about the pioneer who had to keep his gun with him at all times in anticipation of unexpected Indian attack? He might have died from an arrow through his chest but he never died of a heart attack. In those days a man walked the earth as a *man* until his dying days, not merely existing as a senile vegetable for years with the arteries to his brain clogged up as so many of our older people do today—of no use to themselves or anyone else and oftentimes a great burden indeed to their loved ones. (It will be wonderful when our oldsters may enjoy the incalculable blessings of real "life in the years" instead of mere senile "years in the life".) Lest anyone try to discredit the above statement by saying "But the life-expectancy back then was only 40 years and no one lived long enough to develop the modern 'degenerative diseases' ", let us once again consider those uncomfortable things called "facts". The great increase in life-expectancy in the U.S. during the last several decades has been largely a statistical phenomenon. This increased life-expectancy is mainly related to greatly reduced *infant* mortality. The cold, revealing figures show that if you are 50 years old right now your total life-expectancy is only a few *months* more than your grandfather's when he was 50 years old in 1900! There have always been considerable numbers of truly old people alive in any time in history, in all countries of the world.

Next we should consider the possibility of a deficit of some essential substance in the body. It seems most unlikely that there would be a nutritional deficit responsible for widespread gross diseases confined to areas under the influence of modern Western civilization. Quite the opposite would be more logical.

Finally one is obligated to consider the possibility of the presence of foreign substances or the presence in abnormally large quantities of otherwise harmless substances in the body as a cause of disease ("poisoning"). The cholesterol theory of atherogenesis falls within this category —the concept of the presence of abnormally large quantities of fats in the body being responsible for disease. I have already debunked the cholesterol theory, but what about the validity of the above general idea as applied to other substances?

We are now living in an era unique in human existence. Each and every year many dozens of totally new chemical compounds are being introduced into our environment and inevitably into our bodies. The bodies of men throughout the history of human existence have been exposed to chemicals in the environment (mainly from foodstuffs, but also in some cases from pollution of air or water) which were not needed at best and possibly severely damaging—really poisons. But throughout the millennia through the general process of adaptation by means of survival of the fit the bodies of men have developed enzyme systems to handle most common natural poisons, at least to some significant degree. Such adaptation of men and animals in general takes countless hundreds of centuries—but as mentioned above, in the last few decades chemicals totally alien to living organisms (many of which must be frankly considered as poisons) have been and are at present being introduced in incredibly increasing amounts. Therefore when one is studying disease entities which are of very recent origin (e.g. the atheromatous diseases) one must give prime consideration to the possibility of such disease being the result of poisoning—a reaction to a chemical new to the organism and to which it has not had the time or ability to adapt.

If heart attacks and strokes are indeed due to a form of insidious poisoning, where are we to look for the source? There are only a few ways in which chemical substances, including those which must be classified as exogenous poisons, can enter the body. They are: directly through the skin, via the air we breathe, or by means of substances ingested into the alimentary tract, i.e. food or water.

This involves truly a myriad of possibilities and it is easy to become confused and misled. But as alluded to a while back there have been made certain observations of extreme importance, which point the way to the ultimate truth. Others have ignored or denied the importance or even the genuine existence of these findings because they lead away from the generally accepted theories. It is sad that we are in the era of "group think" and procedure by committee. As a great educator once put it, "Could 'Hamlet' have been written by a committee? Creative ideas—spring from individuals."

Just what are these all-important observations and to what do they point? Because the author believes it may be of some future interest, the original observation which led to this new theory of atherogenesis about to be expounded will be singled out and mentioned first. The absolute validity of similar future *in vitro* experiments as regards to extrapolation to living systems is of little importance. What was important is that it provided the nidus for a whole new stream of thought.

This was the seemingly insignificant and widely known fact that in the dairy industry very tenacious, yellowish deposits build up on milk utensils washed in certain kinds of germicidal solutions. It was evident that some chemical in the *water* used to wash the utensils reacted with milk or some component of milk to produce a deposit. The biological analogy should be obvious.

Have there been any experiments reported in the medical literature relating water and elements contained in it to coronary heart disease? It has been shown most conclusively that the harder (higher concentrations of ionized minerals) the drinking water the *less* the incidence of coronary heart disease; and that hardness of drinking water is related to no other known diseases *except* heart and vessel disease!

By now I assume that the reader will have taken the not-too-subtle hints and come to the conclusion that there must be something in *drinking water* that is the culprit. In one short, succinct sentence the cause of atherosclerosis and resulting heart attacks and strokes is none other than the ubiquitous *CHLORINE* in our drinking water!

It is, in my opinion, certainly one of the greatest paradoxes of recorded history that one of the very same public health measures which have been primarily responsible for the great increase in statistical life expectancy in the Western world should also unsuspectedly be responsible for many of the chronic disorders of later life.

It should now be manifestly apparent why no medical scientist has ever even for a fleeting moment entertained the truth. Chlorine is a classical "sacred cow" of modern medical science. Is it conceivable that something of such obvious and wonderful utility, so widely used, and with no apparent acute side effects could be responsible for heart attacks and strokes? Is it not incredibly difficult to believe that millions upon millions of dollars and incalculable hours of time have been poured down the drain by thousands of medical research workers busily and eruditely engaged in finding answers to the wrong questions? And although the two chemicals are totally unrelated, anyone opposing the use of chlorine would be immediately disregarded as an anti-fluoridation-type "nut". What medical scientist would dare risk tarnishing his reputation and grant-collecting future by espousing a theory which would automatically be tainted in the popular mind (due to the popular misconception nearly equating chlor*ine* with fluo*ride* and the smear campaigns against anti-fluoridationists)?

This is, it may be recalled, not the first time in modern medical history that a presumably innocuous chemical material has been indicted as an agent of serious disease. Some years ago blindness caused by a disease known as retrolental fibroplasia was distressingly common among children, especially those born prematurely. Only some years after this form of blindness had become prevalent was it discovered that high concentrations of life-supporting oxygen in incubators caused the disease. Today new cases of this form of blindness are totally unknown; unfortunately for the victims of that era of ignorance the blindness is irreversible.

You will, of course, now want to know on what basis can I make the sweeping statement that *chlorine* in drinking water is the greatest crippler and killer of modern times, i.e. the prime causative agent of atherosclerosis and its end results, the heart attack and stroke.

Let us retrace our steps and see if this new chlorine theory will explain satisfactorily the same facts used in tearing apart the classical cholesterol theory.

Great emphasis was put previously on the concept that the heart attack was essentially an unknown entity before about 1920 and did not become of sufficient significance to affect mortality statistics until about 1930. How does this correlate with the use of chlorine in drinking water? Experimental use of chlorine to "purify" water supplies began in the late 1890's. Chlorination gained relatively wide acceptance in the second decade of this century and in the third decade (1920's) it was found that satisfactory killing of organisms was dependent upon a *residual* of chlorine in the water above the amount necessary to react with organic impurities. When it is remembered that evidence of clinical disease from atherosclerosis takes 10-20 years to develop, it becomes evident that there is a correlation between the introduction and widespread application of chlorination of water supplies and the origin and increasing incidence of heart attacks that is exceedingly difficult to explain away.

In light of the chlorine theory just presented we can now understand why there were no heart attacks in England during the last century despite a significant portion of the population consuming diets high in fat. We can understand why the Eskimos, whose diet is composed in the main of highly saturated animal fats, are immune to coronary heart attacks and other manifestations of atherosclerosis. Why heart attacks have been totally unknown among the Chinese

51

and most all primitive peoples. Why the inhabitants of Roseto, Pennsylvania have no heart attacks unless they move to another community. And we can understand why the gerbil shows no tendency to atherosclerosis.

There was no chlorinated drinking water in England during the last century. Eskimos may consume huge quantities of dietary fats, but their drinking water is pure melted snow. Chlorine in the drinking water is unknown among primitive peoples. The Chinese are basically a poor ignorant race who spread their sewage on the ground and get worms in their guts from drinking contaminated water and eating filthy food. We in the Western world are more civilized —we take our sewage and dump it into our rivers. We then drain it into our water supply, strain it and inject chlorine into it. We don't get worms in our guts but we sure do get something else! The inhabitants of Roseto drink water straight from flowing mountain springs, but when they move to the big city and drink chlorinated water like all the other city dwellers they are subject to the same retribution.

Although not mentioned before: the Japanese who normally have a very low rate of heart attacks are no different from other people when they move to Hawaii—and drink chlorinated water; the Masai tribesmen of Kenya have almost no heart disease although they eat at least as much cholesterol as most Americans—but drink no chlorinated water; coronary heart disease is unknown among a group of 500 poor Irish farm workers studied by famed Dr. Paul Dudley White while being widespread among their chlorine-drinking brothers in the United States. And contrary to popular belief, high level business executives (supposedly under much stress) have a statistically lower incidence of heart attacks than their subordinates. In America when an executive reaches the highest echelons not only does he receive a key to his own "washroom", but while at the office drinks non-chlorinated bottled water.

The lowly gerbil drinks no water at all, instead manufacturing all the water it needs from the dry food it eats, and therefore escapes the end results of chronic chlorine poisoning.

We can understand even the intriguing, apparently unrelated and sometimes apparently facetious facts and correlations presented before. The documented lower incidence of coronary heart disease in areas with hard water could possibly be explained by postulating chemical reactions between free chlorine, an extremely active chemical, and the ions which cause hardness of water resulting in biologically innocuous chlorides. (This is not to exclude the possibility of some other, more complex biological mode of action of the hard water ions.) Even the strange and apparently facetious correlation between the number of telephones per unit of population and the cardiac death rate can be explained quite logically and seriously on the basis that chlorinated drinking water and telephones are both products of modern civilization, both became widespread in the early decades of this century and both are most prevalent in urbanized areas.

There is one other finding that we should not forget to mention. During the Korean War autopsies were performed on otherwise apparently healthy soldiers killed in battle. In an article in the Journal of the American Medical Association it was reported that among the soldiers whose average age was 22.1 years over 75% showed some gross evidence of coronary arteriosclerosis. These results have been widely discussed with the usual conclusion being that coronary artery disease is far more common and extensive than previously suspected, especially in young men. In light of my new theory I most strongly question this conclusion. If you ask any man who served in that war he will tell you that the water in Korea for our soldiers was so heavily chlorinated for sanitary reasons that it was almost undrinkable. Incidentally, discussions with Korean physicians reveal that heart attacks are almost totally unknown among their fellow countrymen. Koreans drink un-"purified" water. (Recent similar postmortem studies performed on U.S. casualties in Vietnam have shown an even higher incidence of coronary disease. Being a miserably hot country the average GI cannot avoid drinking large quantities of incredibly highly chlorinated water. The water made available for soldiers is required to contain a *minimum* of 5 parts per million of chlorine residue. But because only a minimum standard is required by regulations ofténtimes the water is chlorinated at a level several *times* that, levels quite comparable to that in the animal experiments to be described. "Results" on the soldiers seem to be the same as in my experiments. Apparently there is a direct *causal* correlation between the amount of chlorine ingested and the speed and degree of development of atherosclerosis!)

It is very interesting that clinical material (specimens obtained at operation or autopsy) has shown atherosclerosis in synthetic vascular grafts (artificial Dacron fabric arteries) in humans. One cannot help but think of the analogy between this and the deposition of milkstone on smooth rubber or metal surfaces. In both we have the flow of fat and cholesterol-containing fluids over surfaces in the presence of chlorine, with resultant surface deposits. If this analogy holds true, it would be powerful evidence in favor of the so-called "encrustation" theory of the mechanics of atheroma formation, the main differences seen in the artery lining being due to the body's reaction to the deposit. One should note that the encrustation theory offered no explanation for the basic etiology of atherosclerosis. All it did was to try to explain where the deposit came from, not *why* it formed.

Some of the previous examples suggest strongly that dietary consumption of fats bears little relationship to the development of atherosclerosis. This is *not* to say, however, that there is no connection whatsoever between dietary fats, hypercholesterolemia (high blood cholesterol levels) and atherogenesis. To avoid later misinterpretation let me emphasize that atherogenesis involves a system of multiple etiology. This is to say that such factors as diet, exercise, smoking habits, etc. *are* of some significance under appropriate circumstances.

In any system of multiple etiology the primary agent (in atherogenesis the primary agent is chlorine) is only one cause. But it must be an *essential* cause. For example, it would be inconceivable for tuberculosis to occur in the absence of the tubercle bacillus. Nevertheless it is generally recognized that there are multiple other causes than the primary agent (the tubercle bacillus) in the production of clinical tuberculosis. Thus the primary agent must be an essential cause (tuberculosis cannot occur under any circumstances whatsoever in the absence of the bacillus) though not necessarily a sufficient cause (clinical tuberculosis may not occur even in the presence of the bacillus, e.g. tuberculin positive

skin reaction in a clinically healthy person).
Likewise in the process of atherogenesis chlorine
is the essential agent (atherosclerosis cannot
occur to a clinically significant degree in the absence of chlorine regardless of diet and other
contributing factors) though it is not necessarily
a sufficient cause (e.g. normal premenopausal
women do not develop atherosclerosis even if
they are exposed to chlorine and the other factors).

CHAPTER 8

MEDICAL RESEARCH IN THE U.S.— A RACKET?

No radical and new theory, especially one of the immediate and practical importance of that presented in this book, is complete without at least some preliminary convincing experimental proofs. But before we immerse ourselves in this somewhat technical aspect let me air a most unnecessary and even dishonest situation which has permeated the very guts of American scientific endeavor. Our universities and medical schools are thoroughly infiltrated with professional research parasites who are interested in only two things—money and prestige. Medical research has been elevated to the status of a demigod in America. And make no mistake, as a result of this it has entered the realm of financial big business, with all the inherent potential immorality and evilness of the same. Medical research is big business in terms of yearly monetary expenditures—millions upon millions of dollars are spent each year by what amounts to essentially a university monopoly, run by firmly-entrenched, self-designated "research specialists". And what have been the results? In 1966, for example, the National Cancer Institute had spent about $500-million (that's right, half-a-billion dollars) in testing just 170 drugs without any results applicable to the common types of cancers. It makes one wonder if these researchers really do want to find a cure; or if they are primarily concerned with perpetuating their gravy train. And this type of situation is the rule rather than the exception!

What is wrong? It really boils down to another example of the possibility of big money and prestige corrupting absolutely. Make no mistake, I am not condemning medical research *per se* by any means, but rather what has happened to much medical research in America. No longer does a scientist think out a possible solution to a problem and then go to the laboratory to obtain proof supporting his hypothesis. Today with all kinds of monies available the prestige- and money-hungry researchers (and by implication, the prestige- and money-hungry universities and medical schools) obtain financial grants and then go directly to the laboratory hoping that lightning will strike from heaven above and they will "happen" upon some finding of scientific interest or, miracle of miracles, even something of scientific *importance!*

Certain university medical centers have realized the extent of their involvement in this financial big business to the point where they have hired full-time public relations men to tout their researchers. Witness the rash of extravagant and unwarranted claims concerning organ (e.g. heart) transplants! A good PR man can be worth millions of dollars each year to an "up-to-date" medical research center.

CHAPTER 9

THE EXPERIMENTAL PROOF OF THE CHLORINE THEORY

In contrast to the situation just mentioned, the proofs about to be offered are scientific experiments purposely planned to offer direct and straightforward support to or refutation of a previously well thought-out theory.

While it should be understood that the results of animal experiments cannot ever be extrapolated directly to apply to human beings without the ever-present risk of error due to inherent biological differences (human beings are not dogs or rabbits or chickens; remember that Thalidomide was "proven" to be free from deformity-causing side effects by experiments with dogs), there are many instances where if due attention is given to possible sources of error the final conclusions may be quite real and useful. As can be readily realized, there are many important medical experiments which cannot ever be performed on human subjects, at least in a society where we do give lip-service to the inherent worth of human beings and human dignity. Thus we must turn to animal experimentation.

Significant atherosclerosis is almost totally unknown among wild animals, just as it is among primitive peoples. Under special conditions, however, animals of certain species have been made to develop at least some of the early plaques of atherosclerosis in their arteries. With rabbits it is necessary to give the bunnies diets high in added cholesterol (quite obviously a grossly abnormal situation for a normally vegetarian animal). To use the dog a diet high in cholesterol will not do, by itself. In order to induce atherosclerosis in the canine species it is necessary first to destroy the animals' thyroid glands either by surgical removal or by administration of radioactive iodine.

The one other animal species extensively used in the experimental production of atherosclerosis is the lowly chicken. Some researchers maintain that the chicken is prone to the *spontaneous* development of lesions of atherosclerosis, i.e. even when given an apparently normal diet. As will soon be shown this is not true at all, but nevertheless the chicken has been shown to be an excellent experimental species for the purpose at hand. Indeed, it has been widely accepted that the chicken is as good as or better than any other animal for research in atherosclerosis in procedures designed to help analyze pathogenesis and/or therapeutic value of different regimens and substances.

With this knowledge in mind the author set up a controlled experimental situation using the chicken as the experimental species.

There were two general phases to the proof of the chlorine theory. In the first, 100 day-old cockerels were divided into two groups of 50 each. With the understanding in mind that what was to be proven was that chlorine is the *essential* cause of atherosclerosis though not necessarily a sufficient cause, the two groups were set up with known contributory causes acting on both, the only difference being the presence of chlorine—the presumed essential cause—in the drinking water and mash (food) of the experimental group and absent from the food and water of the controls.

The male of the species was chosen in the knowledge that just like human beings it is the male which is primarily susceptible to the development of atherosclerosis. All other experiments on atherosclerosis using the chicken which have been reported in the medical literature have used cockerels for the same reason.

Both groups were placed on a cooked mash consisting of about a 1:1 mixture of corn and oat meals with about 5% low-priced oleomargarine added. Pure distilled water was used exclusively. Chlorine was added to the drinking water and

THE CHARLES E. BLACK CLINICAL LABORATORY
112 West Allegan Street
Lansing 8, Michigan
TELEPHONE IV 2-2044

CHARLES E. BLACK, M.D.
DIRECTOR

PATHOLOGY
AND
DIAGNOSTIC SERVICES

Chicken Aorta

Number: OA-167-67

Date: July 28, 1967

GROSS EXAMINATION:

The previously opened aorta submitted shows a typical raised yellow atheromatous plaque of the aorta. The entire specimen is submitted for microscopic examination.

MICROSCOPIC EXAMINATION:

The sections show a large typical appearing atheromatous plaque of the aorta. The plaque involves the entire wall of the aorta. Areas of liquefaction necrosis and degenerative change are noted in one area of this plaque. No calcium deposits are noted. The adventitia of the aorta shows active inflammatory changes, chiefly lymphocytes.

DIAGNOSIS:

1. Atherosclerosis of the aorta of the chicken with typical atheroma formation. An intimal plaque is characterized by the focal deposition of lipids in the subendothelial connective tissue of the intima. Early, the atheroma is rich in lipid and filled with a soft pink staining grumous material.

The plaque may become fibrotic or calcified. It may ulcerate into the lumen. The intact, or more often the ulcerated plaque predisposes to mural thrombosis. Atherosclerosis, then, causes marked deformity, narrowing and occlusion of arteries. It is responsible for ischemia, atrophy and infarction of dependent structures. As focal endothelial injuries, the atheromas weaken arterial walls and potentiate rupture or aneurysmal dilatation.

2. Atherosclerosis is the most common variant of arteriosclerosis. On this basis, unless the other forms of arteriosclerosis are specifically designated, the term arteriosclerosis is used as synonymous with atherosclerosis. As mentioned above, atherosclerosis is characterized by the formation of atheromas, "focal intimal lipid deposits particularly cholesterol." One theory involves the concept that injury to the endothelial lining of arteries underly the abnormal accumulations of lipid within the intima. It has been shown experimentally that a variety of macro-molecular substances may injure the endothelium and increase its permeability to lipo-proteins. Atheromas tend to occur in arteries at points of stress. Atheromas likewise tend to occur around the mouths of vascular branches and bi-furcations including other points of stress. It is believed that atherogenesis is related to the diet of man and particularly to the level of lipid or cholesterol intact. The plaques contain granular acidophilic, lipid-rich debris and crystalized needle-like spicules of cholesterol. A low grade inflammatory reaction is accompanied by the accumulation of a scant number of lymphocytes, such as is observed in this specimen.

3. Active aortitis involving mainly the adventitial outer, coat.

Charles E. Black, M.D.

CEB/dlk.

67

mash of the experimental group in the form of
chlorine bleach (disinfectant), about one-third
teaspoonful per quart of water. This highly chlorinated water was first given to the experimental
group at twelve weeks of age.

The results were nothing short of spectacular!
Within three weeks there were grossly observa-
ble effects on both appearance and behavior.
The experimental group became lethargic, hud-
dling in corners except at feeding time. Their
feathers became frayed and dirty and the cockerels walked around with their wings hunched
up, their feathers fluffed up as if they were always cold (the experiment was performed in an
unheated barn in winter), their pale combs
drooping. This appearance is most suggestive of
symptoms resulting from clogging up of the
micro-circulation.

Meanwhile the control group was the epitome
of vigorous health. They were much larger in
size than the experimental group, active, quar-
relsome, vigorous appearing with smooth, clean,
shiny feathers and bright combs held up erectly.

No less remarkable was the gross appearance
of the aortas. The abdominal aorta (the place
where atherosclerosis is known to occur in
chickens) of all of the cockerels dying after four
months were carefully examined. In more than
95% of the experimental group grossly visible
thick yellow plaques of atherosclerosis protrud-
ing into the lumens were discovered! These

chickens were noted to have an extremely high apparently spontaneous death rate and common findings on examination of the carcasses were hemorrhage into the lungs and enlarged hearts. Although no blood pressure readings were taken, these findings are suggestive of gross arterial hypertension.

At seven months there were so few experimental chickens remaining alive that the survivors were sacrificed, with identical findings. At the same time one-third of the apparently healthy control group was also sacrificed with not one abnormal aorta found!

. Although these results seemed conclusive, it was decided to repeat the procedures by taking the healthy control animals of the first experiment and dividing them also into an experimental group receiving chlorine and a control group. Once again the roosters receiving chlorine showed fantastic gross changes in appearance and behavior within three weeks. The first change noted was a remarkable paleness of the combs. Instead of bright fiery red the combs became nearly orange in color and soon began to droop. These changes were shortly followed by the ones described for the original experimental group and as expected gross atheromas of the aortas were found on examination within three months.

To summarize, in both experimental groups gross changes in appearance and behavior, most likely explicable by postulating obstruction of the micro-circulation, were evident within a few weeks in the fowls receiving chlorine; followed by gross atherosclerotic lesions of the aortas evident within a couple of months. Control groups treated in an identical manner except for the absence of chlorine remained healthy and vigorous, grew well, and showed no evidence of either atherosclerosis of the aorta or symptoms of possible obstruction of the micro-circulation.

As a side comment, it should be mentioned that the so-called "spontaneous avian (chicken) atherosclerosis" reported in the literature was not spontaneous at all. These experiments, as well as most others on dogs and rabbits, were performed in urban university centers using city water which inadvertently contained chlorine. It has been noted that animals in our zoos are starting to show evidence of atherosclerosis. Once again the culprit is the ubiquitous chlorinated city water.

CHAPTER 10

PRACTICAL SUGGESTIONS

About now the reader of this book is getting more than a little concerned about the water he has been drinking and will drink in the future. Go to the water tap and get yourself a big glass of water. Now hold it up and look at it. What does it mean to you? Perhaps you are a little angry—it is the insidious poison contained in that very glass of water which has been responsible for so much suffering and death of modern times. More men died of coronary heart attacks alone in the last two years in the United States than have been killed in our many wars since our country was founded. Have you ever seen an active, vitally alive human being reduced to a state little more than that of an inanimate vegetable by a stroke?

But rather than being upset, you should be over-joyed. Soon the coronary heart attack and stroke should be no more—mere uneasy memories of a blind and ignorant past. Ever since the conquering of infectious diseases earlier in this century by the use of public health measures (primarily sanitation) and in more recent years the use of antibiotics, the chronic diseases of the aging have come to the fore. But now two of the important "degenerative" diseases have been conquered. Now our citizens may truly enjoy the joys of "life in the years".

What are you going to do about that glass of water? Let me re-emphasize that the development of atherosclerosis is a slow process under ordinary circumstances (e.g. ordinary city water; the heavily chlorinated water given to our combat troops fighting in the stinkholes of the world is another and much more serious matter). There is no need to panic—for most persons the drinking of chlorinated city water for a few more weeks or months will be of little significance.

But what *should* you do now? You must insist, to all persons in positions of power and by all means possible, that chlorination of all water supplies be stopped permanently as soon as is humanly possible.

What can be used to replace chlorination as a means of "purification" of water? While there are several distinct possibilities, e.g. other chemicals (but would these show some terrible side effects in 20 years?), the process most likely to replace chlorination is the passing of thin sheets of water under ultra-violet lights. It has long been known that water flowing through streams and rivers is purified by nature using a combination of friendly bacteria and sunlight, which contains ultra-violet rays. In a similar manner it has been found that pathogenic bacteria in water passed under strong ultra-violet lights are killed, most probably by minute amounts of free oxygen released from the water by the light rays. In addition, water treated in such a manner will kill or impede bacteria added subsequent to the treatment.

Now, a most vitally important point must be considered. Is the use of ultra-violet light systems a practical possibility from an economic standpoint? Does it cost too much to be feasible? Most wonderfully, recent developments in the field of purification of water by means of ultra-violet light have brought down the cost of treatment to no more and potentially less than by means of the deadly chlorine. In addition, with the use of ultra-violet light there is no chemical taste or odor to the water, no storage or mixing of dangerous chemicals, no long storage period required and a new treatment unit is not any more expensive to install than many other types of units. The ultra-violet treatment unit is reasonably compact, is of high capacity and requires a minimum of maintenance.

While at the present ultra-violet light systems seem the most practical alternative to chlorination, one should not exclude the possibility of future development of other economically feasible treatments. For one, ozonation has been proposed. And I am certain that once the American inventive genius is brought to bear upon this problem other procedures will be forthcoming.

Although, as just emphasized, the average person has little to fear in drinking chlorinated water for a few more weeks or months—until the tremendous change-over in method of treating water supplies can be implemented—certain groups of high-risk persons should avoid such water immediately. These are any persons who have had heart attacks or strokes in the past, persons with signs of peripheral atherosclerosis such as victims of intermittent claudication, diabetics, hypertensives, persons with known high levels of blood serum cholesterol and perhaps all men over 40 especially if they are of a muscular build and smoke cigarettes.

What should these groups of people do for drinking water? Perhaps the best water to use if obtainable is pure hard water from a deep, health department inspected well. A second choice is commercially available "spring mineral water" if it can be positively ascertained that no chlorine has been used in sterilization of the carboys. A poor third choice is commercially available distilled water. I have strong suspicions that sometimes chlorine-containing antiseptic solutions are used in cleaning the carboys used for distilled water; secondly, distilled water is devoid of "hardness", i.e. minerals which have been shown to have a protective effect versus chlorine which may be inadvertently ingested; finally, unless it is aerated distilled water tastes terrible.

A temporary but in some circumstances very useful expedient is to bring tap water to a boil and then let it cool. Heating water drives off the chlorine. Hot drinks, e.g. tea, coffee, cocoa, herb teas, etc. are also quite acceptable from the standpoint of freedom from chlorine.

CHAPTER 11

IS EXISTING ATHEROSCLEROSIS REVERSIBLE?

In addition to the avoidance of chlorine ingestion, any high risk person should quite logically avoid assiduously all of the secondary contributing factors mentioned before, i.e. cigarette smoking, overweight, excessively fatty foods, etc.

Although some excellent animal experiments reported in the medical literature have suggested that if a definite cause of atherosclerosis could be found the process could not just be stopped but actually reversed, so far no experiments to prove or disprove this contention have been performed.

There seems to be no reason why the early changes of atherosclerosis should not be nearly completely reversible, depending of course on the total avoidance of chlorine. However there exists the possibility that the prognosis for advanced lesions may not be so good in all cases.

In many disease processes there exists a point beyond which irreversible anatomic changes have taken place. For example, in alcoholic cirrhosis of the liver once the liver cells start to become replaced by scar tissue the course of the disease is inexorably downhill.

The pathogenesis (exact mechanism of development of a disease) of atherosclerosis remains obscure. There does exist the possibility of a heart attack or stroke in a person with far advanced disease even after cessation of chlorine ingestion. Hence the recommendation that high risk persons should avoid any possible contributing factors.

I realize that these last few paragraphs seem to refute some of the more positive statements made previously. Seemingly I proclaim the end of all heart attacks and strokes, only to admit that they may still occur to some extent for a while after the implementation of my suggestions. I am just being careful in my claims. The atherosclerotic process *may* be totally reversible after all. On the other hand, the avoidance of chlorine is primarily a *preventive* measure. While I have unequivocably stated that avoidance of chlorine ingestion will prevent the development of atherosclerosis of clinically significant degree, I do not make the definite claim of "cure" (reversal) of established atherosclerosis.

Certainly no one ever condemned the Salk polio vaccine merely because it had no beneficial effect on a case of clinical polio. Polio was conquered *through prevention*. Likewise with the avoidance of chlorine to prevent the development of atherosclerosis and its resulting clinical syndromes of heart attack and stroke. No one should condemn my theory merely because of the possibility of development of a clinical syndrome in far advanced cases of existing atherosclerotic disease. Heart attacks and the major forms of stroke have been conquered *through a means of prevention!*

CHAPTER 12

INSIDIOUS CHEMICAL POISONING AND THE FUTURE

As I bring this book to a close may I end on a somewhat pessimistic, but honest, note. As mentioned in some detail before, our environment and truly the whole world is becoming increasingly contaminated with new chemicals, the ultimate effects on living organisms including man being almost totally unknown. Most of these substances are at least in some degree cumulative poisons. The ghastly effects of the drug Thalidomide in producing malformed children was overlooked, obvious though it was, for several years. The fantastic long-range effects of chlorine ingestion were not in the least even fleetingly suspected for more than two-thirds of a century. How many other potentially harmful chemical compounds are we being exposed to every day? A cursory glance at the labels in any supermarket is enough to make one shudder with horror. We worry about atomic fallout, but the chemical "fallout" in our environment may be of vastly greater significance to the human race. Alas, the chemical pollution of our environment is a very profitable business. . . .

CHAPTER 13

THE AUTHOR vs. THE ESTABLISHMENT

In the form of a sort of postscript let me say that I fully expect to encounter vast quantities of hostility from professional ignorance and jealousy. In the best of times organized medicine has been known to oppose almost all revolutionary advances. This holds throughout history —witness the scorn heaped upon Drs. Simmelweiss and Holmes in the last century when they insisted that doctors themselves were spreading childbed fever by refusing to wash their filthy hands. Or more recently the disgraceful resistance to the adoption of Sister Kenney's therapy for acute poliomyelitis (later adopted and expanded on—and made "proper" by calling it "physical therapy"). Sister Kenney was a mere nurse who did not spend hundreds of thousands of dollars "researching". And a history of the opposition to the use of anesthesia makes fascinating, if horrifying, reading. These are but a few examples of a long, embarrassing list.

I realize that in this book I have stepped on many feet. There is probably nothing more sacred to modern medicine than that mystical thing called "medical research", which I have debunked in part, at least the dishonest way in which much of it is now practiced. This is more than a matter of principle—a lot of very influential persons in many important places and positions are dependent financially on medical research funds. Inadvertently some will suffer from a monetary standpoint from comments made in this book. In desperation, like a drowning man, when forced to acknowledge my findings these selfsame persons will challenge my logic, my proofs, my conclusions. As there is no place where it hurts more to be hit than in the pocketbook, I fully expect my professional qualifications, my purposes and even my basic integrity to be viciously slighted.

To all of this may I answer in advance: who I am, what I stand for, or anything else connected with me is of little importance. What is important is the health, welfare and lives of millions of persons. I readily admit that if you will work hard enough at it you may possibly be able to find minor flaws in certain arguments, processes of logic, procedures and so on. But nothing, I repeat *nothing*, can negate the incontrovertible fact that the basic cause of atherosclerosis and resulting clinical entities such as heart attacks and the most common forms of stroke is *CHLO-RINE*—the chlorine contained in processed drinking water!

The fallacious cholesterol theory of atherosclerosis has been accepted for many years on the basis of most tenuous evidence. In retrospect we see that many most significant facts, especially those showing that atherosclerosis is almost exclusively a disease of modern Western civilization, have been consciously or unconsciously ignored because they would cast serious doubt on the only semi-plausible theory presented until the present.

On the other hand we see that my chlorine theory accounts for not only all the widely accepted facts concerning atherosclerosis but all the others alluded to previously. Even if I could not produce a single iota of experimental proof to substantiate my claims, one would have to give very serious attention to my ideas on the basis of logical explanation of known facts alone.

But I *do* have experimental evidence to back me up. And my results have been some of the most notable ever presented in an entirely original book on a medical subject.

If at this point you are unconvinced enough to question the basic validity of my conclusions, let me make the following statement: I hereby make an unqualified challenge to all concerned to repeat my proofs—and expand on them. (The basic proof can be duplicated for less than $100. You won't even need a research grant.) *After* you have done this you may criticize as you see fit.

In closing I want to parry one other possible criticism. There will be some who will insist that presenting this fantastic concept of the deadliness of chlorine in producing heart attacks and strokes to the general public without large scale scientific proofs performed in many independent laboratories is "premature". To this just let me ask is it "premature" to delay and let people die when they can live? My responsibilities as a physician are to my patients and society at large, not to certain special interest groups concerned about their own prestige and pocketbooks.

FUTURE AVENUES OF RESEARCH— UNANSWERED QUESTIONS

Many think that given enough money for medical research any disease can be licked, the time needed being inversely related to the quantity of funds made available. They proudly point to the conquering of poliomyelitis to support this assumption. (The large "foundations" have purposely fostered this misconception, for obvious selfish reasons.)

Unfortunately, this line of thinking is totally fallacious. With polio the breakthrough in thinking (i.e. the theory behind immunization as exemplified by smallpox vaccination) had been completed over 150 years before. Conquering of polio, because no new and original thinking was needed, was indeed a simple matter of money. Enough money allowed multiple trial-and-error experiments to find 1. an acceptable medium wherein to grow the virus, 2. a proper method of attenuation of the virus thus produced to create a useable vaccine.

In contrast, any medical research situation where the breakthrough in thinking has not been accomplished is doomed to failure or at least most mediocre results. How can one find a cure for cancer when no two researchers can really agree as to what cancer even *is*? (Yet brazen stories of imminent success are periodically fed to the news media to keep the flow of donations coming.)

But as of the publication of this book no one can deny that the breakthrough on atherosclerosis, heart attacks and strokes is an established fact. At long last further research can proceed on a rational basis with reasonable expectation of significant results from money expended.

Although this book explains in some detail the etiology (causation) of atherosclerosis as being the direct result of chlorine ingestion, the exact *mechanism* of development of the plaques remains obscure. We now realize that atherosclerosis cannot occur to a clinically significant degree in the absence of chlorine, but as yet have no idea as to the exact way chlorine acts in the body to produce the disease manifestations. Is it a straightforward process where chlorine causes a simple deposition of cholesterol in the intima of arteries? Or a vastly complex mechanism involving multiple interactions of the body's whole biochemical system? These questions are of great interest because they could provide a clue as to whether existing disease is reversible and/or possibly even a method could be developed to promote such reversal.

Another question in my mind is the existence and significance of obstruction of the micro-circulation proposed in this book. The chicken experiments certainly suggest most strongly this component of the action of chlorine in the body.

Conservative medical science acknowledges the existence of a disease entity called arteriosclerotic heart disease (oftentimes abbreviated ASHD) which results in a weak, enlarged heart predisposed to arrhythmias and failure. This seems to be a disease quite separate from classical coronary heart disease. Is this ASHD a result of impairment of the micro-circulation caused by chlorine, an effect quite distinct from the chlorine-induced atherosclerotic plaque formation?

Is senility of the aged really a combination disease with a single etiology, i.e. chlorine? Is the blood flow to the brain impaired not only by the development of atherosclerotic plaques in the arteries feeding the brain but also by a direct impairment of the micro-circulation of the brain itself, a second and distinct result of chlorine ingestion.

Other unanswered questions include the mystery of why and how premenopausal women are protected from the deleterious effects of chlorine and the exact mechanism of the partially protective action of hard water.

The results of honest research on the concepts just proposed should prove to be *most* interesting and useful.

YOUR DUTY AS AN INTERESTED CITIZEN

"The reception accorded to Jenner's work (the introduction of vaccination against smallpox) was the same as that usually accorded to great humane innovations. A few people received it with great acclaim, a somewhat greater number opposed it violently, and the vast majority were indifferent."

- "Devils, Drugs and Doctors"
by Howard W. Haggard, M.D.

The time has come for *action* against the greatest killer America has ever known—chlorine. The author has done his part—making available the unadulterated facts, pulling no punches in the process.

The only way the necessary campaign against the killer chlorine can get off the ground is for you, the reader, to get to work on it.

Write your Representatives and Senators! *Demand* that this book and its message be given proper consideration. Insist that your Congressmen read this whole detailed book and then act in behalf of the health of our nation.

Millions of your tax dollars have been spent on medical research on heart disease since the basic concept contained in this book was released in January, 1967. Yet not one cent has been used to confirm my experiments and conclusions; or even in an attempt to refute them.

It's your tax money that has been paying for the existing fruitless research. And it's your tax money that pays for the *un*chlorinated bottled water all high level officials in Washington drink exclusively.

It's up to *you* to demand that the message contained in this book is ignored no longer.

241